Strategic Studies Institute
and
U.S. Army War College Press

STRENGTHENING STATEHOOD CAPABILITIES FOR SUCCESSFUL TRANSITIONS IN THE MIDDLE EAST/NORTH AFRICA REGION

Mohammed El-Katiri

April 2015

The United States Army War College

The United States Army War College educates and develops leaders for service at the strategic level while advancing knowledge in the global application of Landpower.

The purpose of the United States Army War College is to produce graduates who are skilled critical thinkers and complex problem solvers. Concurrently, it is our duty to the U.S. Army to also act as a "think factory" for commanders and civilian leaders at the strategic level worldwide and routinely engage in discourse and debate concerning the role of ground forces in achieving national security objectives.

The Strategic Studies Institute publishes national security and strategic research and analysis to influence policy debate and bridge the gap between military and academia.

The Center for Strategic Leadership and Development contributes to the education of world class senior leaders, develops expert knowledge, and provides solutions to strategic Army issues affecting the national security community.

The Peacekeeping and Stability Operations Institute provides subject matter expertise, technical review, and writing expertise to agencies that develop stability operations concepts and doctrines.

The Senior Leader Development and Resiliency program supports the United States Army War College's lines of effort to educate strategic leaders and provide well-being education and support by developing self-awareness through leader feedback and leader resiliency.

The School of Strategic Landpower develops strategic leaders by providing a strong foundation of wisdom grounded in mastery of the profession of arms, and by serving as a crucible for educating future leaders in the analysis, evaluation, and refinement of professional expertise in war, strategy, operations, national security, resource management, and responsible command.

The U.S. Army Heritage and Education Center acquires, conserves, and exhibits historical materials for use to support the U.S. Army, educate an international audience, and honor Soldiers—past and present.

STRATEGIC STUDIES INSTITUTE

The Strategic Studies Institute (SSI) is part of the U.S. Army War College and is the strategic-level study agent for issues related to national security and military strategy with emphasis on geostrategic analysis.

The mission of SSI is to use independent analysis to conduct strategic studies that develop policy recommendations on:

- Strategy, planning, and policy for joint and combined employment of military forces;
- Regional strategic appraisals;
- The nature of land warfare;
- Matters affecting the Army's future;
- The concepts, philosophy, and theory of strategy; and,
- Other issues of importance to the leadership of the Army.

Studies produced by civilian and military analysts concern topics having strategic implications for the Army, the Department of Defense, and the larger national security community.

In addition to its studies, SSI publishes special reports on topics of special or immediate interest. These include edited proceedings of conferences and topically oriented roundtables, expanded trip reports, and quick-reaction responses to senior Army leaders.

The Institute provides a valuable analytical capability within the Army to address strategic and other issues in support of Army participation in national security policy formulation.

*****

Comments pertaining to this report are invited and should be forwarded to: Director, Strategic Studies Institute and U.S. Army War College Press, U.S. Army War College, 47 Ashburn Drive, Carlisle, PA 17013-5010.

*****

This manuscript was funded by the U.S. Army War College External Research Associates Program. Information on this program is available on our website, *www.StrategicStudiesInstitute.army.mil*, at the Opportunities tab.

*****

*****

The Strategic Studies Institute and U.S. Army War College Press publishes a monthly email newsletter to update the national security community on the research of our analysts, recent and forthcoming publications, and upcoming conferences sponsored by the Institute. Each newsletter also provides a strategic commentary by one of our research analysts. If you are interested in receiving this newsletter, please subscribe on the SSI website at *www.StrategicStudiesInstitute.army.mil/newsletter.*

ISBN 1-58487-671-9

FOREWORD

The political transitions that followed revolutions in Egypt, Libya, and Tunisia are by no means complete. They have gone through challenging phases, characterized by political volatility and tension. These challenges and uncertainties have a profound impact on the three countries' economies and the ability of their new political elites to build credibility and political legitimacy.

Each of the three countries studied in this monograph has followed a distinct path, with its own opportunities and limitations. Policy reforms during each transition have reflected individual political, security, and social realities. However, one common lesson learned from their trajectories is that their challenges are exacerbated by a deficit of human and institutional capabilities to design and implement political and economic reforms.

In this monograph, Dr. El-Katiri focuses on the variety of challenges that face reform attempts in the post revolution Arab countries. He examines underlying factors that have prevented new political elites and post-revolution institutions from successfully implementing essential reforms, and how these failures have affected the building of their political legitimacy. He concludes with a series of specific policy recommendations, with the aim of preventing further deterioration to the detriment of U.S. interests.

The Strategic Studies Institute is pleased to offer this analysis to provide readers with information for further debate on this topic.

DOUGLAS C. LOVELACE, JR.
Director
Strategic Studies Institute and
U.S. Army War College Press

ABOUT THE AUTHOR

MOHAMMED EL-KATIRI is a Director of MENA Insight, a political risk consultancy that focuses on the Middle East and North Africa, and a Senior Research Analyst at the United Kingdom's (UK) Conflict Studies Research Centre (CSRC). Before joining CSRC, Dr. El-Katiri was a Research Fellow at the UK Defence Academy, and later served as a Political Risk Analyst at Eurasia Group and as a Senior Researcher at the Hague Institute for Global Justice. With more than 10 years of regional experience, Dr. El-Katiri focuses his research interests on political and economic security in North Africa and the Gulf Cooperation Council (GCC) states, as well as on North African relations with the European Union and security policies around the Mediterranean. Dr. El-Katiri has published numerous internal and external UK Defence Academy reports, including reference papers on national and regional security issues in the GCC and the Mediterranean region, and a monograph on the Algerian national oil and gas company, Sonatrach. In addition, his publications include commentaries and research papers in a range of languages for research organizations in the UK and Europe, as well as various newspaper articles in and about the region. Dr. El-Katiri is a frequent media commentator, including for the British Broadcasting Company, the *Financial Times*, and *Al-Jazeera*.

SUMMARY

Despite achievements made during the past 4 years since Tunisia, Egypt, and Libya overthrew their former dictatorial rulers, these societies still face major governance, security, and economic challenges. These challenges vary from country to country depending on their specific circumstances, including the intensity and duration of the uprising, and the prevailing economic and political situations before and after the upheaval. Libya's security and political situation in particular remain chaotic. Very limited progress has been made on rebuilding basic state institutions or solving legal problems related to restoring justice for individuals and communities.

The inability of the new political elites to meet the political and economic expectations of their populations has undermined their political legitimacy which has had a significant impact on a variety of economic sectors in all three countries, and on their public finances. For instance, Tunisia's and Egypt's public finances have significantly worsened over the last 4 years, resulting in large fiscal deficits. Lack of financial resources has limited the ability of both countries to tackle the structural problems of their economies.

However, failures to implement reforms and resolve socio-economic problems cannot be attributed solely to lack of financial resources. The capacity of post-revolution political elites and institutions was constrained by a skills deficit among bureaucratic cadres. Politicians and civil servants have had to learn from scratch the complexity of the challenges that face their societies in the post-conflict era, in economic recovery, justice, security, and more. The new era requires innovative thinking to address economic difficulties, draft new legislation, and establish new institutions.

This monograph assesses the post-revolution challenges faced by these three countries, and measures progress made in reaching them. It reviews their implications for the broader region and for U.S. interests, and makes a number of policy recommendations for U.S. decisionmakers.

STRENGTHENING STATEHOOD CAPABILITIES FOR SUCCESSFUL TRANSITIONS IN THE MIDDLE EAST/NORTH AFRICA REGION

INTRODUCTION

Political transitions that follow revolutions or wars are inevitably challenging phases, characterized by political volatility and tensions. Transitions are periods that require the rebuilding of important political institutions and skillful handling of a variety of acute political, security and socio-economic challenges. Transitions are also phases that often entail the reform of security apparatuses, reform of the justice system, writing new or introducing substantial amendments to existing constitutions, and in some instances the re-engineering of the state governance system with the establishment of new structures to ensure the distribution of political power and wealth. On the economic side, governments have to launch policies to foster economic growth and regain the trust of local and international investors. It is a complex and long process. In addition, these policy areas often have to be tackled simultaneously, given their interconnectedness, particularly at a time when popular expectations are high and every move of the government is under intense scrutiny.

Several countries in the Middle East and North Africa (MENA) region have experienced all of these multiple challenges since 2011, following the overthrow of Arab leaders who had ruled their respective countries for decades. The triumph of protests in Tunisia and Egypt, deposing Zine El Abidine Ben Ali and Hosni Mubarak, inspired other popular revolts to force their authoritarian rulers out in Libya and Yemen, or to achieve greater political concessions in Algeria and Morocco.

This monograph takes the case studies of Egypt, Tunisia, and Libya to examine the scope for ensuring successful political transitions in the MENA region's current unsettled state. The post-revolution political elites in these three countries are struggling to strengthen their legitimacy in the eyes of the public, and their attempts to gather support for the new political institutions and reforms have made limited progress. Despite their aspirations, the past 4 years have been characterized by political volatility and continuous instability.

Rebuilding challenges differ from one country to another according to individual circumstances. Specific needs depend on the intensity and duration of the uprising, and the prevailing economic and political situations before and after the main upheaval. For instance, Libya, a country that experienced approximately 6 months of civil war, had the urgent matter of returning and restoring justice to displaced communities and individuals. The war caused massive displacement of communities and residents of cities both within Libya and abroad. Furthermore, in the specific case of Libya, the fall of the Qadhafi regime left the country without functioning security institutions but with dozens of rival armed groups, and politics and security thus became intertwined. In Egypt and Tunisia, the popular revolts were violent but did not escalate into civil war, and internal security has remained in the hands and under the control of the Ministries of Defense and of the Interior in both countries.

Addressing these challenges requires a wide range of skills, knowledge, institutional capacity, and coordination abilities at top echelons of government. These requirements hardly exist in any of the three countries analyzed in this article at this point in time. The rebuilding tasks are beyond their capacities alone, and experience of previous international transitions that followed conflict or revolutions shows that external assistance is essential. But at the same time, international aid has been reduced by the austerity policies in place in most developed economies. The cuts in government budgets across Western countries are affecting their willingness and ability to provide economic development, justice, and security assistance in places that need it most.

This monograph does not aim to offer a comparative analysis between the three countries, but rather looks at their different trajectories in the ongoing transitional phase to build new political institutions and governance. The research builds on the author's working experience with officials and academics from the region within a range of training and capacity-building projects.

The first section will focus on why reforms are an inherent part of development in the post-revolution Arab countries. The second section looks at different reform attempts in key areas initiated over the past 3 years which have failed to make any successful progress. The third section examines a range of factors that constrained progress on a number of reform projects across the three countries, and notes that the weaknesses of these governments cannot only be attributed to the current political context, but are instead an outcome of decades of bad staffing and recruitment in both the public and private sectors. The fourth section argues that capability building needs differ greatly between the three countries, and that tailored responses are required based on a thorough assessment of these needs.

THE LOGIC OF REFORM IN TRANSITION

The upheavals that swept the MENA region after December 2010 were driven by an irrepressible desire for change in both the political and economic spheres and the restoration of justice. Millions across the Arab countries protested against rampant corruption, unfair economic conditions, and the lack of political freedoms and dignity. The calls for the departure of their respective leaders or regimes were a reflection of a desire for a prosperous future within inclusive economic and political systems. Thus, the new political elites and institutions had limited choice but to be responsive to the demands and aspirations of the broad base of their populations and adopt ambitious reforms.

Each of the three countries described here has had its distinct path, opportunities, and limitations. Policy reforms during their transitions have reflected individual political, security, and social realities. While there was indisputable regime change in the case of Libya, the other two countries have seen more nuanced political scenarios. The 2011 revolutions in Egypt and Tunisia saw the ousting of key political figures and the fall of governments, but did not result in an uprooting of regimes. In both countries, the security apparatus that was an important pillar of the ousted regimes remained unchanged. The dismissing of senior officials in Tunisia did not result in a substantial change of loyalties, beliefs, and *modi operandi* of the security organizations.[1] These organizations and their unofficial affiliations with established business networks constitute what became commonly known as the "deep state," which has shaped political events during the transition. Although they were not perfect democracies, both of these countries had political parties and established bureaucracies, with levels of efficiency which were acceptable by regional standards.

Libya, by contrast, represents a more complex and difficult environment. The country is plagued by continuous instability, and interim governments have lacked the capacity or commitment to exercise leadership in developing state institutions. Lack of qualified personnel in what is left of the Libyan bureaucracy, and a limited academic environment, constrain rapid and effective progress in the rebuilding of state institutions.

Legitimacy and Constitution-Building.

The post-armed conflict or post-revolution governments of Egypt, Tunisia, and Libya are confronted with a range of challenges which must be addressed in order to build their legitimacy and overcome the *sequelae* of the instability that led to the political upheaval in the first place. This crucial transitional political phase requires the enactment of a variety of new laws and establishment of new institutions to deal with tasks which are specific to this period. This has been evident in all three countries, where the new governments have enacted new electoral laws and established special-purpose institutions to draft a new constitution, organize and oversee elections, or introduce reforms. Once constitutions are promulgated, parliaments and governments tend to establish a variety of structures and institutions to implement the constitutional provisions, and this undertaking in turn implies the design and regulation of further new institutional structures.

Scholars of post-conflict societies deem the rewriting of the constitution to be an important step in laying the foundation for the country's democratic transition to accountability, inclusive political participation, rule of law, and protection of human rights. Thus, all changes of government in the Arab countries over the last 4 years have dictated either the drafting of a new constitution where it has never existed, as was the case with Libya, or introducing substantial amendments to the existing ones in the cases of Tunisia and Egypt.

The announcement of the drafting of a new constitution is regarded as a declaration by the new political elites of their intent to enshrine the protection of human rights, democracy, and rule of law principles and institutions. In the absence of a constitution, the interim Libyan National Transitional Council promulgated a Constitutional Declaration in August 2011, setting the governance structure for the transition phase until a new constitution is drafted.[2]

Drafting a new constitution is deemed paramount to building the credibility and legitimacy of new governments, and thus deserves to be prioritized. The drafting of a constitution is not an easy task in post-revolution or post-regime change contexts. It is a process that requires not only skillful handling of the content and wording of the constitution, but also deft management of the process. It is usually an undertaking that is fraught with risks and challenges, especially in a transition period characterized by weak political legitimacy and fierce completion among a variety of political groups. This too has been evident in the cases of Egypt, Tunisia, and Libya. For instance, in Tunisia, which led the political upheaval in the Arab world, it took the National Constituent Assembly 2 years and 3 months to complete the drafting of the new constitution which was approved on January 26, 2014. The process of deliberation was neither smooth nor without controversy. The divergence of views between Islamist and secular parties on a variety of issues reached the level of political crisis on several occasions. The position of religion in political life has been one of the most controversial matters in all three countries, and one that has generated vigorous political debates within the constituent bodies and outside among political parties and civil society.[3] However, the legal debate has to some extent only mirrored the competition in the political arena among political parties, and the limited trust they have in each other.[4] The debates, despite the tensions they sparked, provided opportunities to deepen understanding of a variety of ideals and what they mean to different social and ideological groups within Tunisia.

The drafting of Egypt's constitution under the Muslim Brotherhood's rule was not without controversy either. Secular and leftist forces protested the dominant position Egyptian Islamists have in the Constituent Assembly, with their fears centered on the role of Islamic law, Sharia, as source of legislation.[5] In Libya, the unfortunate position society found itself in with weak or no vital state institutions, and no constitution or constitutional culture among its intelligentsia, complicated progress on drafting the constitution. The process of preparing the country's first constitution since the end of monarchy in 1969 became the victim of the political and security chaos that has dominated Libya over the last 4 years.[6] The power struggle between different ideological, regional, and ethnic armed groups hampered the process of electing a constituent assembly. The General National Congress and different post-Qadhafi governments had to balance competing demands on the representation issue from ethnic minorities such as Amazigh, Touareg, and Tabu, as well as regions, secularists, and Islamists. The process of drafting is still at an early stage since 60 members of the constituent assembly were elected in February 2014.

The drafters of constitutions have to deal with a variety of vital themes ranging from identity matters to the organization of the state structure and the status of its vital political and economic institutions. For example, vital themes that are usually discussed include, but are not limited to, the change from a centralized to a federal state system, the structure and size of parliament, and the judiciary governance system. This is a fundamental project that requires careful study of each of the options and thorough discussion.

Moving away from political aspects associated with constitution making, the process requires the availability of a wide range of resources and skills. Experts usually look at experiences of other countries that have gone through similar transitions and political transformations. But not all of these countries have had access to technical expertise, or were well informed about comparative constitutional literature and experiences.

Restructuring the Security Sector.

Another fundamental area of reconstruction is reform of the security sector. The impetus and extent for reforming the security apparatus in post-regime change differs from one country to another. In some countries, the main driver is to remove remnants of the previous regime, either because of the fear of loyalty to the ousted regime or because of their involvement in abuses of human rights. In others, the reform is driven by a desire to change the culture of their institutions and subject them to more accountability and civilian oversight, which is an important condition for the transition to more representative political systems.

In most cases, these reforms are responsive to popular demands, particularly where these security agencies have been the most visible and detested manifestation of the old authoritarian regimes. This is the case in the three countries studied in this monograph; the security agencies were involved for decades in a variety of repressive policies and events.

Thus, part of the reform entails the adoption of regulations and other measures to introduce mechanisms for oversight of security organizations by the government and parliament or judiciary, as well as the provision of training to existing forces to change their working practices in line with the new accountability and respect for human rights. Other parts of the restructuring, discussed further, involve strengthening their capabilities quantitatively or qualitatively to deal effectively with a range of threats.

Once again, the needs for reform differ across the three countries, depending on the political and security circumstances of each state. Libya remains an extreme case. The fall of the Qadhafi regime was accompanied by the fall of the state security institutions. The current government security institutions lack the capability to maintain order, protect citizens and the country's borders, and critical facilities. The security landscape is further complicated by the existence of a variety of rival well-armed groups, which emerged as a result of the circumstances of the 2011 war. As predicted in an earlier monograph,[7] most military arsenals inherited from the Qadhafi regime and the subsequent civil war remain in their hands.[8] Thus, security sector reform in Libya entails not only the formation of new professional security institutions that are not based on regional, tribal, or ethnic allegiances, but also must include plans to demilitarize armed groups and help the reinsertion of combatants into civilian and security institutions. In fact, this has been a priority for the interim Libya governments since late 2011.

The picture is less complicated in Egypt and Tunisia. Both countries have strong and functioning security apparatuses, with Egypt having the largest armed forces by number of personnel in the Arab world. Their main motivation is primarily a desire to restore the loss of credibility of the security forces in the eyes of the public. Thus, the main objective is to focus on strengthening governance and respect of human rights. In Egypt, the ideal main goal of reform would be to ensure gradual disengagement of the military from politics and economy. Over the last 6 decades, since the military coup in 1952 that saw the ousting of King Farouk, the military has expanded and consolidated its position in both the politics and the economy of Egypt.[9] Egyptian presidents since that time have been military officers, including Mubarak. In Egypt, there is no civilian oversight over the management of defense affairs, and the Ministry of Defense is run by the military. Meanwhile in Tunisia, the reform of the Ministry of Interior's security agencies is seen as an integral part of the wider democratic transition strategy. During the Ben Ali era, different security agencies which fell under the umbrella of the Ministry of Interior were involved in assassination, torture, and other maltreatment for years. The reputation of the police and security forces deteriorated still further as a result of its brutality in quelling protests that led to the departure of Ben Ali, as well as during subsequent protests after January 2011.[10] The gathering of thousands of Tunisian protestors at the time outside the Ministry headquarters in Tunis evidenced how it epitomized regime oppression.[11]

Given the vital role these security agencies play in the functioning of the state and its stability, reform decisions are deemed delicate, and they are rarely launched immediately in the aftermath of a government change. The new political elites might immediately signal the need to rehabilitate the security agencies, but leave any substantial reforms to later stage. Once launched, the reform process tends to be gradual. Pursuing deep reform at an early stage and in a comprehensive way runs the risk of demoralizing and alienating security officers and officials, and their political backers. Furthermore, any attempt to revisit the injustices caused by the security agencies over the years endangers the stability of the political transition.

Economic Reform.

Socio-economic grievances played an important role in sparking the revolutions in these three countries. But the economic situation has deteriorated still further since December 2010. The troubled political transitions, characterized by demonstrations and security disorders, have caused significant economic losses and have brought the economies and public finances of Egypt and Tunisia near to collapse. Instability has affected economic growth, discouraged direct investors, and led to substantial falls in foreign reserves and the rise of budget deficits. In Libya, the 19th largest oil producer in the world, the picture is no different. Oil exports were halted for months during the continued insecurity and thus deprived the Libyan state of its main source of revenues. Oil production and export facilities became the targets of armed rebel groups in their dispute with the Libyan government. The situation remains fluid as the intertwined security and political tensions remain unresolved.

It is unsurprising that the post-revolution governments have opted for a set of policy reforms that attempted to revive the economy, improve the delivery of basic services to their population, and tackle issues of income inequality and regional development disparities. In order to placate public discontent and mark the start of a new political era, interim governments in Tunisia and Egypt introduced measures to promote youth employment and maintained subsidies for food and energy products.

In addition, the interim rulers in the three countries launched initiatives to track billions of dollars stolen by the deposed rulers, their families, and entourage.[12] These initiatives were deemed important for strengthening values of democracy and morality, as well as consolidating trust through demonstration of the discontinuity between the new rulers and deposed ones. The recovery of illegal assets has been prioritized because of its economic significance, particularly in the case of Egypt and Tunisia where the need for funds to shore up dwindling public finances, and the countries' foreign currencies reserves is huge and immediate.

LIMITED SUCCESS IN DEALING WITH CHALLENGES

The political transition that followed the overthrow of the rulers in Egypt, Libya, and Tunisia is still unfinished. No strong and legitimate government or leader has emerged in any of the three countries. The mere fact of holding a number of elections to choose temporary institutions to handle the transition has not led to any significant progress. The post-2011 governments have not yet consolidated their stance on the political scene. The three countries have not had stable governments; either there has been a major reshuffle or an entire change of government as happened in Egypt. Even Tunisia, which has been less politically turbulent compared to Egypt and Libya, experienced a severe political crisis that started early in 2013 and lasted for months. The source of tension was ideological between the ruling Islamist Ennahda party and its secular opposition. In December 2013, after months-long arduous negotiations, a new apolitical government was announced.[13]

In Egypt, the victory of the Muslim Brotherhood in the 2012 legislative and presidential elections did not yield any political stability. On the contrary, the tussle between the remnants of Mubarak regime, mainly from the security apparatus, and the ruling Islamists escalated and led to a military-backed coup in June 2013.[14] The ousting of the elected president, Mohamed Morsi, inaugurated a new phase of unrest that has not yet settled despite the organization of new presidential elections,[15] which resulted in the election of a retired Field Marshal Abdel Fattah El-Sisi. At the time of this writing, sporadic clashes between Muslim Brotherhood supporters and Egyptian forces of order are still taking place.

Libya's transition has seen dramatic cycles of political instability. Since the fall of the former regime, Libya has descended into an era of deep ideological, tribal, and regional tensions and continuous security threats. Following the ousting of the elected Libyan

Prime Minister Ali Zeidan in March 2014, Libya entered a further period of institutional uncertainty that lasted for months. The election of Ahmed Maiteeg as successor was deemed unconstitutional by Libyan liberal political parties and their supporters, and led to a power rivalry between him and caretaker Abdullah al-Thini, who took office temporarily after Zeidan.[16] Insecurity continues to prevail on the ground, with persistent clashes between a variety of armed groups and the assassination of security officers and targeting of politicians. The military operation initiated by renegade general Khalifa Haftar since mid-May 2014, with the advertised aim of defeating extremist groups and bringing stability, has further worsened the security situation and spread the fighting from Benghazi to Western parts of Libya, including Tripoli.[17] After months of fighting, Khaifa Haftar's military coalition has lost several battles and with them the control of Tripoli and Benghazi in August 2014, which are now controlled by Misrata-led forces and Islamists, respectively. At the time of this writing, the country continues to be unstable as fights between Khalifa Haftar's forces and their adversaries have not ceased. The current military polarization is also reflected in the political realm. Libya again entered a period of political confusion in August 2014. The country now has two parliaments, one in the eastern city of Tobruk and another one in Tripoli, and each has named its own prime minister.[18]

What these political events reflect is the inability of the new political elites to win the trust of the great majority of their populations, or to engineer an inclusive transitional political process that enables different political parties to work together. To cultivate trust, the leaderships should articulate the aspirations of the people and the interests of key political stakeholders. This political element is vital for successful and peaceful transitions. But it is not the only factor. New governments also have to meet socio-economic goals. The change of regimes has to result in betterment of the lives of citizens in a sustainable way.

Limited Tangible Deliverables.

The failures in rebuilding trust come as a result of limited success in improving the delivery of basic services. Over the last 3 years, no long-term policy has been launched to deal with the many structural challenges facing these countries. Their governments have been predominantly in fire-fighting mode, reacting to crises and not proactively launching policies to consolidate different aspects of their economies, particularly in areas with potentially major economic and political impact. No policies to reform health services, provide affordable housing, or tackle unemployment have been announced. There is a lack of any detailed plan to create new industries and strengthen their economic diversification, or amend energy and food subsidy systems to consolidate fiscal balances.

The opposite has happened; most of the programs announced have been short term, indicating a deficit of vision among the new leaders. The few populist initiatives which were adopted caused other secondary effects. In the absence of any structural plans, the governments resorted to increasing public spending to placate popular anger. For instance, the decision of the Tunisian government to increase the number of public sector jobs to absorb the number of unemployed graduates did not provide a structural solution to the unemployment issue. The Tunisian government created 34,000 jobs in public service and state-owned companies, an equivalent of fewer than 5 percent of the total unemployed.[19] Meanwhile, these new appointments have added more pressure on the public finances.

Surprisingly, even in countries with a record of remarkable economic reforms such as Tunisia, the pace has slowed down. This is mainly attributed to the focus of the leadership on elections and constitutions. During this period, Tunisia experienced a high turnover of senior civil servants in key ministries and public owned companies. After government reshuffles, new ministers have tended to appoint their trustees in key ministerial departments and public companies.[20] The constant replacements have impacted the capacity of bureaucracy to elaborate and implement complex reforms. These shortcomings have fed into the worries of international investors and lending organizations, and constrained the flow of international capital into Tunisia which has further complicated the country's economic situation.

Transitional Justice.

The establishment of transitional justice for past violations of human rights is important for building trust between the people and the state, as well as for strengthening the social fabric of the three transition countries. Transitional justice, in fact, has become one of the key demands of civil societies in the aftermath of 2011 revolutions for crimes that happened during the rule of ousted leaders or afterwards during the transition, particularly in Libya and Egypt. However, implementation has suffered from lack of capabilities and excessive politicization. In Libya, the National

Transitional Council (NTC), acting under the influence of the revolutionary fervor of its own people and idealist advice of international human rights donors, enacted several regulations to tackle human rights abuses committed under the former regime and during the 2011 war. In 2011, the NTC enacted law 17/2011, establishing a Fact Finding and Reconciliation Commission, a body with a mission to investigate human rights crimes but no mandate to punish them. In May 2012, NTC issued Law 38, granting amnesty to revolutionaries who had committed criminal acts which were dictated by the circumstances of the 2011 war.[21] This law attracted criticism from international human rights organizations and experts as giving a blanket amnesty to all criminals, including perpetuators of crimes against humanity and war crimes.[22]

The enactment of these two laws highlights two main shortcomings. First, it showed an absolute miscomprehension of the sensitivity of such political initiatives and the conditions that they require for success. The implementation of a mechanism of transitional justice in a political context characterized by continuing tensions among a variety of regional and ideological rival groups could further exacerbate conflict. The laws were enacted in the absence of even basic state institutions to oversee and enforce their application. Second, these laws were criticized for being incomplete and lacking specific elements for its application. For instance, Law 17/2011 did not specify the period and type of crimes to be covered by the Commission's investigation work. Article 1 of the law opted for an elastic approach by stating that the investigation will cover all infringements of human rights. The commission also lacked the equipment and skilled officials to conduct its mission, such as forensic experts and materials.

The laws were adopted with no accompanying strategy to update the country's criminal code and procedures to be able to deal with war crimes, or to provide training to the Libyan judges and prosecutors, most of whom lacked any knowledge or experience of how to deal with the atrocities of the previous regime.[23] Given these shortcomings, the Libyan government, with the assistance of subject-matter experts and international organizations, had to issue a new comprehensive law which was promulgated by the General National Congress in December 2013. Law 29/2013, on transitional justice, abrogated the previous texts and established a new Fact Finding Commission with a specific mandate, and detailed regulation of its functioning and relations with other state institutions.[24]

The promulgation of the Political Exclusion Law in May 2013, which disqualifies officials and officers who served under the former regime from playing any role in the new political and public institutions, is another example of chaotic approach to transitional justice adopted by Libyan interim governments.[25] The law disqualifies all people associated with former regime, even if they did not commit any criminal act while serving under Qadhafi rule. The exclusion of the process from any judicial review undermines efforts to build a new political regime that is based on rule of law. The law was unpopular not only among the former civil servants who are affected by it, but also among civil society and political activists. This lustration has undermined what is already a fragile political scene, and is depriving the state of thousands of experienced professionals such as the military officers, ministers, and judges.[26]

Disarmament and Reintegration of Militias.

Another vital policy area that has failed to achieve its main goals is the disarmament, demobilization and reintegration (DDR) of Libyan rebels. The DDR program was badly designed and executed, leading to a debacle which has contributed to the continuing instability in Libya. The Libyan authorities, with their international advisors, lacked detailed mapping of the ideological, political, and economic motives of different armed militias. The planning process did not pay attention to the difficult context in which some of their programs would be implemented. As a result, the efforts of the Commission for Warriors' Affairs (CWA) to reintegrate rebels into civilian life or in the police and defense forces achieved limited success. A considerable number of large militias and their members did not join the DDR programs. They are still unwilling to hand over their weapons or enroll into official state security structures. The incentives and programs for reintegration have simply not been attractive, or even suitable for the majority of rebels.

Besides pursuing political goals, large militias are generating substantial income for themselves from the provision of protection to state-owned and private assets. Some are involved in the smuggling of goods, drugs, weapons, and people across neighboring borders. In some ways, the formation of a national army and police forces are able to provide security across the Libyan cities and regions, as well as protect its borders, would be contrary to their interests. This also explains a series of murders and assassination attempts against key senior military officers struggling to build a national army. The lack of a diversified or functioning economy that is able to offer real employment opportunities has also impeded sustainable reintegration of ex-fighters into civil life. With the scarcity of employment opportunities, it is likely that some beneficiaries of the reintegration program will eventually be recruited back into armed militias.

The absence of well-paid job opportunities offered by the private sector has further limited the ability of the Libyan authorities to convince rebels to hand over their weapons. The education level of many rebels does not qualify them to take over civil service jobs or work in state-owned companies, but those jobs in the private sector which do not require a high level of skill are filled by foreign workers. The government remains the main employer of Libyan nationals, and there is a limit to what it can absorb.

Programs to promote entrepreneurship among armed militiamen, through the provision of training and funds to create micro and small enterprises, were doomed to fail because of continued violence and the absence of a suitable business environment. One of the few successful programs launched by CWA has been a scholarship program for post-graduate studies abroad, which is intended to benefit 18,000 ex-fighters by 2015.[27]

ROOTS OF TRANSITION DEBACLES

The transition phase in the three countries has been characterized by fragility, which hampered progress on a variety of essential reforms. This was a period marked by intense political jockeying between different parties over power, and the type and scope of possible reform. The new leaderships have been confronted with a complex chessboard to manage in order to avoid a return to violent instability, or at least mitigate its intensity when it arrives. The three countries have gone through periods of violent riots, assassinations, and terrorist attacks to varying degrees. The tense political environment has diverted valuable resources and distracted attention from much-needed reforms and institutional rebuilding initiatives. Naturally, the more politically sensitive reforms are even more influenced by the power balance between political stakeholders.

The difficulty of implementing police reform in Egypt during the Muslim Brotherhood rule is a case in point. Attempts at gradual reform of the police forces were faced with resistance from officers. In early-2013, hundreds of police forces went on strike in different Egyptian cities, asking for the ousting of Interior Minister Muhammad Ibrahim and his appointees in the Ministry of Interior.[28] Meanwhile, in Libya, the unsettled security and political situation has influenced the pace of rebuilding the country's armed forces as well as limiting the type of schemes which can be adopted to achieve this goal. The Libyan authorities have tried a range of approaches, from co-opting armed militias through paying them wages from the Ministries of Defense and Interior, to creating new security institutions and sending their recruits for training abroad. Even the training of defense forces in several North Atlantic Treaty Organization (NATO) member states, which was the subject of high hopes, is expected to make limited change despite all efforts to harmonize the training program among the host NATO partners. Ana Gomez, European Parliament Standing Rapporteur for Libya, observed after her visit to Tripoli during the first week of October 2013:

> The rebuilding of capable armed forces and police in Libya requires a sustained and coordinated international effort of assistance that will demand several years and commitment from Libya and its partners. That assistance requires training of forces in the country and not just token offers of training abroad, as offered by some EU MS [member states].[29]

The outcome of this hybrid strategy was a disappointing failure, and came at a high cost for Libyan elected politicians in terms of financial costs and credibility.[30] The incapacity of the government was palpable when it failed in November 2013 to respond to citizens' demands to force or negotiate a retreat of armed militias. Tripoli then witnessed general strikes and daily marches of Libyan citizens calling for the government's military and police to take over the security of the capital.[31] The difficulty the Libyan government experiences in trying to mobilize what military forces are left is due to the politicization of the military command. Senior Libyan generals have taken sides in the power tussle between Islamist and non-Islamist actors in the parliament and government.

Difficulties in Political Crisis-Management.

Lack of progress cannot, however, be attributed only to fragile political and security factors; other influencing factors have been in play. Lack of knowledge and experience among the new political leaders in dealing with the complications of transition, and a shortage of experienced professionals and financial resources remain the main drivers of difficult and disorderly political trajectories.

The course of events over the last 4 years shows that these societies were institutionally unprepared to deal with sudden and complex change. The largely unplanned protests and ousting of presidents took everyone by surprise, both in the MENA region and internationally. The protests were driven mainly by youth and without any ideological underpinning. The interim and newly-elected governments had no plans and no understanding of how to deal with a sudden change of the political scene, with all the specific challenges that comes with it. The situation was further complicated by a dearth of think tanks or academia to provide intellectual assistance and professional advice on developing a vision to govern the transition or refine some of its key elements.

The deposed ruling elites lacked detailed contingency plans to deal with some of the political challenges that Egypt and Tunisia faced after their revolutions and Libya after its civil war in 2011, such as constitutional changes and economic reforms to strengthen budget balances and ensure better distribution of wealth. Even if such plans existed, they were likely kept in the possession of a tight circle of trusted collaborators in Egypt and Tunisia. This was certainly the case in Libya, where the ideas for political and economic reforms which were promoted by Qadhafi's son, Seif Al-Islam, were unlikely to have reached a comprehensive or advanced stage.[32] The newly elected Islamist parties in Egypt and Tunisia had no experience of being in power and were not acquainted with the subtleties of governing a state. In many instances they were not able to evaluate the advice they received from international organizations and consultants related to post-conflict issues.

Neither did history provide guidance to the current rulers in tackling the challenges they are facing. After independence from European colonialism, the political leaders of these countries had certainly overcome hurdles to assert their leadership and build their state institutions, but the political and security environments in which this was achieved were far less complex than those of today. Political elites then did not have to act in an environment dominated by the internet, social media, and rising popular expectations that are difficult to meet.

The lack of experience was manifested in the reluctance of governments to initiate reforms. In some instances, the politicians' insecurity was reflected in not making decisions. In many instances, their reactions were the opposite of what the times of extraordinary crisis, in fact, required. This was the predicament of elected Libyan Prime Minister Ali Zeidan, who was widely considered weak and lacking charisma to assert his authority.[33] Successful transition requires a decisive and strong government to push through difficult reforms and attempt to break up vested interests. All the previously mentioned limitations constrained the performance of these post-revolution governments, and their ability to rebuild a new political order and meet the expectations of their voters.

Insufficient progress during transition cannot be attributed solely to elected ministers; instead, it reflects a wider failure to develop quality human capital over previous decades. This issue also concerns other elements of these MENA societies. The shortage of talented bureaucrats, professionals, and businessmen remains the main constraint that has limited the ability of the new political leadership to design new economic and political governance, in a context that is characterized by political volatility. Presidential and ministerial cabinets did not have well-established experts with the skills and experience to develop and oversee the implementation of strategies in a variety of policy areas, including in primary areas such as economic and foreign policy.

It goes without saying that the quality of the economic and bureaucratic intelligentsia remains a decisive factor that determines the success of a transition country in adapting to and shaping challenging events. But those qualified and experienced individuals who might have been able to advise new governments had generally been ousted as part of each new government's policy to cut out key figures from former regimes. The parallel option of drawing more on qualified nationals in academia, industry, or from abroad, is typically not used either, as the decades of closed-circle political elitist culture in these countries renders the outreach to a skilled citizenry intrinsically difficult even for new political parties. With typically no existing expertise of their own within the government's underlying bureaucratic layers, these newly elected governments hence took office armed with little more than vaguely formulated goals of doing everything differently from the previous regimes, and little practical planning and implementation capacity.

Learning Politics from Scratch.

The post-revolution and war eras have showed a deficit in the skills of the bureaucracy cadres across the region. Politicians and civil servants have had to learn from scratch the complexity of challenges that face their societies in a post-conflict era in areas of economic recovery, justice, and security. The new era requires new legislations to tackle issues of transitional justice, the drafting and negotiations of constitutions, and training for judges to deal with crimes that were committed in the very special circumstances of civil war. It requires innovative thinking to address economic difficulties. In Libya, for instance, judges investigating human rights violations were constrained by lack of legal texts, and most importantly lack of experience of how to deal with such cases either as prosecutors or judges. Another example is the lack of procedures or expertise for tracing the wealth and properties of former rulers in Egypt, Libya, and Tunisia. After the fall of the three regimes in 2011, investigators were constrained by their lack of experience in understanding complex legal and financial structures in international investments.[34]

Furthermore, there are other causes that are structural and emanate from decades of poor management of human capital, nepotism, and corruption. The failure of education systems to provide young generations with skills or entrepreneurial abilities relevant to the labor market has jeopardized attempts to placate public anger. There are variations in population sizes among the three countries—Egypt is the most populous country in the Arab world with more than 80 million people—but one shared characteristic is the large proportion of their citizenry which is made up of young people. Lack of skills and abilities is preventing a significant portion of the population from benefiting from the existing economic opportunities. In countries like Egypt and Tunisia, for instance, there are plenty of opportunities for micro and small enterprises in the tourism and agriculture sectors that have not been seized by young people.

It follows that when the governments of these three countries attempt to diversify their economies into new industries and markets, they will face further constraints. Moving to new industries requires specific knowledge and management skills which are scarce in the current labor market. But it is not only a lack of specific business education and training that has led to a deficit of entrepreneurial initiative. Other challenges which inhibit the creation of new businesses include an inability to access finance, lack of business networks, and prohibitively expensive commercial properties.

Inability to Resolve Structural Economic Problems.

The problems with high youth unemployment in these countries are significant and will have a persistent impact on their societies' stability and prosperity, if not addressed. Given that unemployment has always been an economic challenge in these countries, the next logical question would be what is different now. In fact, other socio-economic developments over the last few years have worsened the situation. First, the number of unemployed youth with degrees has increased without any concurrent significant expansion in job opportunities. Second, the role of families in supporting young members is waning as a consequence of inflationary pressures that have weakened their abilities to provide assistance. Third, lifestyle change and the spread of consumerism are adding more pressures onto the societies of these three countries and widening the gap between rich and poor.

The unavailability to these governments of financial resources to fund development projects and policy reforms has presented a major challenge to their ambitious reform programs. This is especially the case for Egypt and Tunisia, which do not have the levels of revenues enjoyed by neighboring hydrocarbon-rich countries. The paucity of public finances has hampered their ability to increase wages, maintain subsidies for basic products, or embark on major infrastructure projects. These states all have urgent needs to launch major economic reforms either to diversify their sources of energy, or build other physical infrastructures that are essential to increase the competitiveness of their economies and to attract more direct investments. But the large-scale investments that these projects require are not available locally.

The only available solution is to approach international private and public sources of financing and create investment and financial schemes on a project-by-project basis. This path does not come without its political complexities and expenses. The engineering and negotiation of complex investment and financial schemes is time-consuming, and require highly skilled experts to manage them on the part of the governments. The experience of Egypt's negotiations with the International Monetary Fund (IMF) to obtain a loan of U.S.$4.8 billion illustrates the laboriousness of dealing with international financiers. At the time of writing, negotiations have lasted almost 3 years with no conclusion, as a result of disagreement over conditions proposed by IMF officials who requested a range of reform measures to reduce the budget deficit, including reduction of subsidies and increased taxation. Successive Egyptian governments have found the IMF measures politically nonviable at this fragile political and economic juncture.[35]

SHAPING RESPONSES TO THE CAPABILITIES CHALLENGE

The three countries have been weakened by the 2011 revolutions and the continuous instability that has dominated their political scenes since then. The adverse repercussions of the unstable political and security situation on the economies and fiscal balances of these countries are now well documented.[36] It has hindered the recovery of tourism in Egypt and Tunisia and led to a significant drop in oil production and export in Libya. To restore peace and prosperity, the three countries all need to build trust in their new governments. This predominantly entails enabling the different states to perform their vital functions as providers of basic services to their citizens, including the provision of security and justice, and also to play an important role in reviving the economy.

There are wide variations in capability deficit between the three countries. Common features that exist should not mislead us into assuming that their needs are identical. Each nation has its unique features and circumstances, and needs vary across the three countries depending on their current levels of political stability, economic development, and political history. Some countries are in a more precarious situation than others. For instance, Libya is a country with almost no functioning state institutions. "One size fits all" therefore is not a viable approach to deliver successful capacity-building assistance. No single solution will fit all these countries or even a specific sector across the region. Solutions have to be individualized and tailored to respond to the needs of each country, or each sector within one country. Thus, there is no substitute for conducting thorough needs assessments for each country to identify ways and means to improve the capabilities of existing institutions and structures, and to build new ones.

Capabilities development is a broad concept covering a wide range of activities intended to strengthen the ability of people and organizations in a variety of sectors across government, private sector, and business to run their affairs effectively. This can be a complex and lengthy process. This monograph is primarily concerned with the institutions and areas that are of highest impact on the entire process of institutional rebuilding during transition, and which in the cases of the three countries under review need to be prioritized without delay.

Priority for Security.

There may be no area that should be given more priority than the rebuilding and capability strengthening of the security apparatuses in the three countries to ensure that order and respect for law are preconditions for any other political and economic reforms. The post-revolution security landscape within the three countries has transformed to an extent which only a handful of analysts previously would have thought credible.

Tunisia and Egypt remain more stable than Libya and retain functioning military institutions. But an increased threat of terrorism has been evidenced by an escalation in attacks, and also an increase in the number of their nationals participating in Jihadist activity in the Levant and Iraq.

In addition, the deterioration of the situation within Libya, and the inability of the Libyan armed forces to control the country's vast borders, have added to the security concerns of its neighbors. Spillover has included dramatic expansion in cross-border trafficking of weapons, humans, and other illicit products. The proliferation of arms in the border areas has increased the number of terrorist attacks in Libya's neighboring countries. The renewal of fighting in July 2014 once again led to destabilizing flows of refugees and smuggling into Libya's neighbors. The new wave of refugees from Libya has become a particular source of concern to the Tunisian government. The financial repercussions of the huge influx of refugees, who are adding further pressure on Tunisia's public finances,[37] have forced the government to call on the international community for economic support to cover the costs associated with humanitarian assistance, and have increased demand for subsidized products and pressure on the health sector. Other negative spillovers include damaging the reputation of the country as a safe destination for tourists and investors.

Instability in Libya currently dominates regional discussions. Egypt and Tunisia, with other neighboring countries, have held several security meetings at ministerial levels to discuss appropriate measures to handle the cross-border security impact. So far, limited progress has been made in formulating a response to tackle various dimensions of this regional security problem, including within Libya. To complicate matters further, the North African and Sahel regions' potentials for mutual security support are undermined by the continuous disagreement and distrust that characterize relations between its two most militarily capable countries, Algeria and Morocco.

Institution Building.

Another area which requires prioritization is equipping the three governments, preferably at the level of prime ministers or presidential cabinets, with an institutional capability that is able to originate and develop large-scale development projects. This is a capability that is needed to develop visions and provide high quality advice and expertise on a variety of governance, security, and economic development matters, as well as oversee the implementation of critical initiatives for the stability and prosperity of their respective societies. This requires a cadre of well-trained strategists and diplomats from among existing senior civil servants who are able to respond to the change of circumstances within their countries and beyond. This corps must be able to adapt to the new environments characterized by scarcity of resources and high competition, and to pursue new avenues to find innovative solutions to pressing social, legal and political issues, as well as opening new markets and sources of financing for their economies.

It adds to the misfortunes of these societies that their transitions come at a time of major shifts in the world economic and political orders. The developed world is going through deep economic adjustments that have affected its ability and willingness to engage in the stabilization and reconstruction of other countries. Most donors from Organization for Economic Cooperation and Development countries have adopted austerity budgets which involve significant spending cuts on international development. There is a change both of attitude and of perception among citizens of major donor countries. Public opinion is less supportive of increased or ineffective spending abroad at the same time as cuts reduce the services offered at home.[38]

Governments and non-governmental organizations from developed nations at first had shown readiness to assist Tunisia and Egypt with their political and economic reforms. An initiative baptized the "Deauville Partnership" was launched in May 2011 during the annual G8 summit held in the French city of Deauville. The main component of the initiative was a package of multibillion dollar loans offered to several Arab countries including Tunisia, Libya, and Egypt between 2011-13.[39]

Three years after the launch of the Deauville Partnership, little progress has been made in alleviating the economic pressures on both the governments and the peoples of Egypt and Tunisia. One of the criticisms leveled at the international financial institutions involved in the channeling of these loans concerned their inability to adapt their procedures to provide innovative ideas that were suited to the specific socio-economic contexts of the transition countries. The proposed conditions to accompany the loans remained unchanged from those in place over the previous 3 decades. International financial institutions did provide funds for major infrastructure projects in electricity generation, and funds to finance small and medium enterprises, which will contribute to the long-term growth of the Egyptian and Tunisian economies. However, there has been a dearth of initiatives launched or sponsored by these multilateral lending institutions which are aimed at improving living standards and reducing social discontent. Greater priority should be given to reforms that will strengthen the legitimacy of the government as a provider of services and producer of opportunities to all its citizens, including the most vulnerable members of society.

The current situation demands severe rationalization of development expenditure. It follows that improved institutional capability at the top echelons of power in these countries in transition is all the more essential, in order to ensure superior coordination of rebuilding efforts and initiatives at a national level, as well as to make better use of the limited assistance offered by international development partners. In addition, one of the shortcomings of reconstruction efforts to date has been a lack of coordination between different policy fields so as to ensure that policy responses are comprehensive. Improved capacity will thus also enhance the ability of these governments to embark on a variety of urgent political and economic reforms without delay.

CONCLUSIONS

Progress in stabilization, reform, and building new institutions in the three MENA countries under review has certainly not been spectacular. But while the political situation remains tense in all three states, the achievements and debacles vary from one country to another. Tunisia has made substantial progress on the political front compared to other countries affected by the Arab Spring upheavals. But at the same time, there has been slow progress in addressing the original economic causes of the upheavals. The political and economic transition in Egypt remains fragile, and the new government still has a long way to consolidate its legitimacy and restore sustainable economic stability. In Libya, the transition has been chaotic, marked by continuing conflict among rival armed groups which has severely impeded the rebuilding of state institutions and the economy and widened dividing lines within society, making Libya a country at serious risk of failure. The renewal of heavy fighting between militias in Tripoli and Benghazi since May 2014 is drawing the country once more to the brink of civil war.

The needs of these societies are various, complex, and interdependent. They range from holding elections, rearranging the justice and security sectors, to reconstructing the economy. Addressing these tremendous tasks is beyond the capability of a state alone, even if it had its own ample financial resources. It follows that efforts to promote post-conflict institutional rebuilding require the collaboration of a range of different governmental and nongovernmental national and international actors. In other words, for these countries to build their legitimacy and achieve sustainable political stability, international partners must provide assistance to enable the new governments to meet their pressing obligations in the short term. This is an investment in the future stability of the entire region.

Trust can dwindle quickly, as has been demonstrated on several occasions over the last 3 years. Failure to deliver on services and to ensure social justice—which entails creating employment and fairer distribution of the country's wealth—has caused riots in all these countries. People were demanding immediate solutions to their grievances. In Tunisia, President Moncef Marzouki has been greeted with the same slogans that were used during the protests that led to the ousting of Ben Ali; "get out" or "people want the fall of the regime." In 2012, the second anniversary of the incident in the Tunisian city of Sidi Bouzid that sparked the revolution in December 2010 saw residents showing their anger and disappointment by attacking politicians with tomatoes and stones.[40] This incident encapsulated the disenchantment with the slow progress that Tunisia had experienced almost 2 years after the ousting of Ben Ali.

The chronic failure of Libya's elected governments to build security institutions and exercise effective authority across the entire territory of the nation is affecting the security of the Maghreb and Sahel regions as a whole. The inability of Libyan government armed forces to control the country's borders is unquestionably a threat to the regional stability, not only in the Maghreb but also beyond. The situation is further aggravated by the absence of regional cooperation mechanisms and responses. The Maghreb countries have not yet managed to build a regional security cooperation forum, partly because of the rivalry between Algeria and Morocco regarding regional leadership; but also because, with the exception of Algeria, these countries have limited human and material resources to carry out intelligence gathering, border patrols, or deploy forces into Libya. This is another reason for international partners, in particular the United States, to maintain and increase engagement with support security sector reform and capability development, in order to reduce further risks of instability and terrorism.

Assisting economic recovery during transition can be highly effective, and reduce the risk of renewed political chaos. Strong economic growth will bolster the nascent political elites' legitimacy by meeting the expectations of a large number of their citizens, and thus stabilize the country. Linkages between unemployment, political discontent, and instability are well documented, and repeated rioting in all three countries over the last 3 years occurs when it becomes evident to the populace that the authorities are unable to meet their expectations for a dignified standard of living.

Similarly, a lack of political stability during transitions frequently coincides with a lack of economic opportunity, which generates a vicious cycle in which religion, ethnicity, or tribe reassumes a pivotal role in people's daily lives as a means of ensuring access to economic means. The reconstruction of a functioning private sector has received little attention in the past 3 years from the international community, which has damaged the ability of these conflict-torn societies to rebuild the sort of economic stability which would foster sustainable political stability and peace.

Given the shared interest in maintaining social order and ensuring smooth transitions toward stability in the region, international and local authorities should engage in designing public policies that respond to private sector concerns and encourage investment. It is essential to understand all factors that hinder the flow of investments and provide a set of measures to mitigate these risks and encourage investments in all sectors, rather than limiting the focus to lucrative natural resources and physical infrastructure markets. More attention should be given to the potential of mobilizing the small and medium sized companies that are currently constrained by lack of resources and expertise. Small and medium sized companies are often unable to handle the complexities and challenges involved in investing capital in high-risk environments. This will not only require regulatory and financial support, but also institutional support to guide and accompany business investments.

POLICY RECOMMENDATIONS

The MENA region is strategically important to the United States and developed countries in political, economic, and security terms. It follows that profound instability in the region continues to pose a challenge to U.S. interests in all of these fields. Resolving this instability requires not only sustained effort and political wisdom from within the region, but also assistance and facilitation from the United States and other partners.

Stability in the Middle East and North Africa is a matter that goes beyond the sole competency of the region's own states and organizations to deal with; it is an international matter. The importance of stability and democracy in the region is not only limited to economic reasons, but also assuring that there are no radical political forces holding power in the country, including, but not limited to, Islamists. Furthermore, more than one of the countries under discussion commands a strategically important location in the Mediterranean.

Strengthening the capabilities of transitional governments and institutions in Arab countries that experienced abrupt change of leadership should continue to be an important part of U.S. Government engagement in the region. U.S. governmental and nongovernmental organizations, including the U.S. military, should continue to prioritize assistance to build capability and strong institutions.

Lessons from the recent past suggest that failure to support political and societal transition effectively increases the vulnerability of internal instability as well as regional and international peace. Such instability can further affect U.S. economic and geostrategic interests in these countries in transition and beyond. But the potential impact on future U.S. security interests of continued instability can be mitigated by leveraging and applying U.S. experience in stabilization, state-building, and institution building, including security sector reform with the assistance of the U.S. Army. These security sector reforms, which constitute an important part of strengthening the new political regimes, also continue to offer excellent cooperation and development opportunities to the U.S. Army, with the potential for establishing and bolstering long-term beneficial relationships with key regional security actors.

The U.S. Government, jointly with other international donors, should accelerate public sector human resource and institutional development in the three countries, through tailored sustainable capacity building initiatives. The core aim is to equip these societies with future cadre of leaders in a variety of policy areas to achieve an expeditious transition. Development of skills should be linked simultaneously with strengthening institutional capabilities—through the establishment of procedures, systems, and provision of equipment—to ensure sustainable retaining of expertise, skills, and knowledge.

Recent events demonstrate that the stability of Libya is vital for the security of the entire Maghreb and Sahel, and for the security of U.S. allies on both shores of the Mediterranean. The U.S. Army can assist other international forces in stabilizing Libya in order to strengthen confidence among legitimate political factions, and thus contribute to the institutional development.

Libya needs effective security institutions, including professional and well-equipped armed forces.

The troubled transition that Libya has experienced was due in large extent to the absence of an effective and nonpoliticized security apparatus. In June 2013, a handful of NATO countries agreed to host Libyan soldiers for training for a few months. Thus, the U.S. military with other international allies have an important role to play in training and equipping the Libyan armed forces far beyond that program so as to foster enduring security sector reform and stability. Any such international engagement should be systematic, comprehensive, and take place on Libyan soil in order to achieve sustainable results. By contrast, the sporadic engagements of U.S. forces to train and assist the nascent Libyan armed forces over the last 2 years remain fruitless and far from achieving the rebuilding of an effective Libyan military force.

There are significant obstacles to further U.S. involvement in security sector reform and military training in-country. In addition to the current chaotic situation in Libya, the United States is challenged by multiple engagements on other important fronts, and a potential deficit of political will for further engagement in degraded security environments. However, this should be balanced against the perception within the region of a moral obligation on the part of the United States to be involved in stabilization after its (albeit limited) involvement in operations to remove the previous regime. Put another way, if European and Arab states were to take on the burden of training and support without U.S. involvement, this would bolster negative perceptions of the United States as an irresponsible and hostile actor, both in Libya and in the region more broadly. At a bare minimum, the United States could coordinate international support, assist with equipping the armed forces, and provide more places for Libyan servicemen in U.S. defense training and education facilities.

Libya also is in need of nonmilitary security rebuilding programs. The U.S. foreign policy and development institutions could increase their engagement with rebuilding effective police and judicial institutions, as these sectors of statehood are crucial for the re-establishment of rule of law and strengthening the credibility of the state in the eyes of its institutions.

In light of increased terrorist threats posed by domestic and regional radical groups, the Tunisian and Egyptian security forces are also in need of further U.S. military assistance in order to strengthen and adapt their existing capabilities to asymmetrical threats. Prolonging U.S. intelligence sharing agreements, coupled with increased training programs and the supply of intelligence gathering materials, will reinforce the capability of each country to confront terrorist threats. The Egyptian and Tunisian transitions remain fragile. A further deterioration of security in either or both countries will further complicate their economic recovery and political transition, with likely instability spillovers beyond their borders.

The struggles of the Arab transition countries cannot entirely be attributed to lack of financial resources. A deficit of vision among the new leaders and a lack of capability to comprehend the current challenges and respond with timely and effective solutions, remain the main handicap to successful transitions. The United States should support and facilitate the establishment of an institutional capability in the countries under discussion, comprised of senior officials who can coordinate and prioritize steps toward achieving transition goals. A body of this kind would not only play a coordinating role, but also provide insightful advice on how to deal with the challenges that the respective governments face now and will face in the future.

ENDNOTES

1. "Tunisie: Lutter Contre L'Impunite, Restaurer La Securite," ("Tunisia: Combatting Impunity, Restoring Security"), Middle East/North Africa Report No. 123, Brussels, Belgium: International Crisis Group, May 9, 2012.

2. Jason Gluck, "Extending Libya's Transitional Period: Capitalizing on the Constitutional Moment," *Peace Brief*, Washington, DC: United States Institute of Peace, September 1, 2011, available from *www.usip.org/publications/extending-libyas-transitional-period-capitalizing-the-constitutional-moment.*

3. Hajer Jeridi, "Tunisie, la place de l'Islam dans la constitution fait débat," ("Tunisia, the Position of Islam within the Constitution Generates Debate"), *Le Monde*, March 1, 2013, available from *www.lemonde.fr/international/article/2012/03/01/tunisie-la-place-de-l-islam-dans-laconstitution-fait-debat_1649916_3210.html.* Yadh Ben Achour, "Religion, révolution et constitution: spécialement d'après l'exemple tunisien," ("Religion, Revolution, and Constitution: The Case of Tunisia"), *Leaders*, November 13, 2012, available from *www.leaders.com.tn/article/religion-revolution-et-constitution-specialement-d-apres-l-exemple-tunisien?id=9831.*

4. Mohammed El-Katiri, *Revival of Political Islam in the Aftermath of the Arab Uprising: Implications for the Region and Beyond*, Carlisle, PA: Strategic Studies Institute, U.S. Army War College, July 2014.

5. Sherif Tarik and Hatem Maher, "Egypt's Constituent Assembly Unveiled Amid Fears Over Islamist Dominance," *Ahramonline*, June 13, 2012, available from *english.ahram.org.eg/NewsContent/1/64/44716/Egypt/Politics-/Egypts-Constituent-Assembly-unveiled-amid-fears-ov.aspx.*

6. Mohamed Eljarh, "One Step Closer to a New Libyan Constitution," *Foreign Policy*, September 26, 2013, available from *transitions.foreignpolicy.com/posts/2013/09/26/one_step_closer_to_a_new_libyan_constitution.*

7. Mohammed El-Katiri, State-Building Challenges in a Post-Revolution Libya, Carlisle, PA: Strategic Studies Institute, U.S. Army War College, October 2012.

8. "Final Report of the Panel of Experts Established Pursuant to Resolution 1972 (2011) Concerning Libya," S/2014/106, New York: United Nations Security Council, February 19, 2014, available from *www.securitycouncilreport.org/atf/cf/%7B65BFCF9B-6D27-4E9C-8CD3-CF6E4FF96FF9%7D/s_2014_106.pdf.*

9. Zeinab Abul-Magd, "The Egyptian Military in Politics and the Economy: Recent History and Current Status," *CMI Insight*, Christopher Michelsen Institute, No. 2, October 2013.

10. "Report of Special Rapporteur on Torture and Other Cruel, Inhuman or Degrading Treatment or Punishment, Juan E. Méndez: Mission to Tunisia," A/HRC/19/61/Add.1, New York: United Nations General Assembly, February 2, 2012, available from *www.univie.ac.at/bimtor/dateien/tunisia_unsrt_2012_report.pdf.*

11. David D. Kirkpatrick, "Tunisia Leader Flees and Prime Minister Claims Power," *The New York Times*, January 14, 2011, available from *www.nytimes.com/2011/01/15/world/africa/15tunis.html?pagewanted=all.*

12. "Report of the Arab Forum on Asset Recovery," Arab Forum on Asset Recovery, September 11-13, 2012, available from *star.worldbank.org/star/sites/star/files/report_arab_forum_on_asset_recovery.pdf.*

13. "En Tunisie, le nouveau gouvernement obtient la confiance" ("In Tunisia, the New Government Gets the Approval"), *Le Monde*, January 30, 2014, available from *www.lemonde.fr/tunisie/article/2014/01/29/tunisie-confiance-du-parlement-au-nouveau-gouvernement_4356033_1466522.html.*

14. Azmi Bishara, "Revolution against Revolution, the Street against the People, and Counter-Revolution," Research Paper, Doha, Qatar: Arab Centre for Research and Policy Studies, September 2013, available from *english.dohainstitute.org/file/get/789010dd-4e5a-4902-8610-cbf77ea22f46.pdf.*

15. Mohamed Morsi was elected in June 2012 after what is considered the first free presidential election following the January 25, 2011, revolution. Under his rule, the country experienced continuous protests and political crises as result of the failure of his government to address the people's demands, and win the trust and backing of the bureaucracy, secular parties, and business elites. Egypt's economy reached critical position as consequence of constant political tensions and instability, with dwindling foreign reserves, and mounting budget deficit. For detailed discussion of the difficulties of Muslim Brotherhood during their rule of Egypt, see Mohammed El-Katiri, *Revival of Political Islam in the Aftermath of the Arab Uprising: Implications for the Region and Beyond,* Carlisle, PA: Strategic Studies Institute, U.S. Army War College, July 2014.

16. Suliman Ali Zway and Kareem Fahim, "Libya Leader Finally Picked but is Quickly Challenged," *The New York Times,* May 4, 2014, available from *www.nytimes.com/2014/05/05/world/middleeast/libya-leader-finally-picked-but-is-quickly-challenged.html?_r=0.*

17. Borzou Daragahi, "Voices from Libya's Armed Struggle," *Financial Times,* July 27, 2014, available from *www.ft.com/cms/s/0/88ca2da4-1585-11e4-ae2e-00144feabdc0.html#axzz38rH2rXYy.*

18. Feras Bosalum and Ulf Leassing, "Rival Second Libyan Assembly Chooses Own PM as Chaos Spreads," *Reuters,* August 26, 2014, available from *uk.reuters.com/article/2014/08/26/uk-libya-security-idUKKBN0GP0NX20140826.*

19. Karim Mejri, "Unemployment in Post-Revolutionary Tunisia: The Public Sector, Object of Desire," *Nawat,* September 20, 2012, available from *nawaat.org/portail/2012/09/20/special-report-unemployment-in-post-revolutionary-tunisia-part-2-the-public-sector-object-of-desire/.*

20. "Tunis: Fièvre des nominations, avant le départ des ministers" ("Tunisia: Nominations Fever before the Departure of Ministers"), *African Manager*, March 3, 2013, available from *www.africanmanager.com/147975.html?pmv_nid=1ll.*

21. "Report to the Secretary-General on the United Nations Support Mission in Libya," S/2012/675, New York: United Nations Security Council, August 30, 2012, available from *unsmil.unmissions.org/Portals/unsmil/Documents/SGReport30August2012.pdf.*

22. Human Rights Watch, "Libya: Amend New Special Procedures Law," *News Release*, May 11, 2012, available from *www.hrw.org/news/2012/05/11/libya-amend-new-special-procedures-law.*

23. Interviews conducted by the author with six Libyan judges and prosecutors in June 2013 during their participation in a training course on transitional justice in The Hague, Netherlands.

24. "Law 29/2013 about Transitional Justice," Tripoli, Libya: General National Congress, available from *www.aladel.gov.ly/main/uploads/sections/459_Transitional_Justice_Law.pdf.*

25. Sami Zaptia, "Political Isolation Law Passed Overwhelmingly," *Libya Herald*, May 5, 2013, available from *www.libyaherald.com/2013/05/05/political-isolation-law-passed-overwhelmingly/.*

26. Mohamed Eljarh, "Isolation Law Harms Libya's Democratic Transition," *Foreign Policy*, May 8, 2013, available from *transitions.foreignpolicy.com/posts/2013/05/08/isolation_law_harms_libya_s_democratic_transition.*

27. Said Farhat, "Libya Tursilu 18 'alf mina Thuwar lidirsa fi lkharij" ("Libya sends 18 Thousand of Revolutionary fighters to Study Abroad"), *Alarab*, April 17, 2014, available from *www.alarab.co.uk/?p=28788.*

28. David D. Kirkpatrick, "Angry at Public and Officials, Security Forces Strike in Egypt," *The New York Times*, March 7, 2013, available from *www.nytimes.com/2013/03/08/world/middleeast/angry-at-public-and-officials-police-strike-in-egypt.html?_r=0.*

29. European Parliament Standing Rapporteur for Libya, "Report on my Latest Visit to Tripoli, Libya: 4-7 October 2013," available from *www.anagomes.eu/PublicDocs/3832ce4c-06f8-430e-ad00-65306af36bfb.pdf*.

30. An intervention by Abdul Rahman Al-Ageli, Security File Coordinator in the Prime Minister's Office, in a virtual seminar organized by Centre for Security Governance. See "Libya: Dealing with the Militias and Advancing Security Sector Reform," *eSeminar Summary report*, No. 1, Centre for Security Governance, January 2014, available from *www.ssrresourcecentre.org/wp-content/uploads/2014/01/eSeminar-1-Summary-Report-08-Jan-14.pdf*.

31. Ghaith Shennib, "Tripoli's Leaders Call for anti-Militia Protests," *Reuters*, November 17, 2013, available from *uk.reuters.com/article/2013/11/17/uk-libya-militias-idUKBRE9AG04F20131117*.

32. George Joffee, "Saif al-Islam: the Whole (Libyan) World is in his Hands?," *Sada*, Washington, DC: Carnegie Endowment for International Peace, January 13, 2010, available from *carnegieendowment.org/2010/01/13/saif-al-islam-whole-libyan-world-in-his-hands/6bal*.

33. Informal interviews conducted by the author with a group of Libyan lawyers and academics in Tripoli, Libya, September 21-26, 2013.

34. Robert F. Worth, "Obstacles Mar Quest for Arab Dictators' Assets," *The New York Times*, June 7, 2012, available from *www.nytimes.com/2012/06/08/world/middleeast/libya-egypt-and-tunisia-try-to-recover-assets.html?pagewanted=all&_r=0*.

35. Rebecca M. Nelson and Jeremy M. Sharp, "Egypt and IMF: Overview and Issues for the Congress," Report for the Congress, Washington, DC: Congressional Research Report, April 29, 2013, available from *fas.org/sgp/crs/mideast/R43053.pdf*.

36. "Arab Countries in Transition: An Update on Economic Outlook and Key Challenges," Washington DC: International Monetary Fund, April 9, 2014, available from *www.imf.org/external/np/pp/eng/2014/040914.pdf*.

37. Raouf Khalsi, "La Tunisie seule, face une perile libyen" ("Tunisia Alone Faces the Libyan Peril"), *Le Temps*, July 31, 2014, available from *www.letemps.com.tn/article/85048/la-tunisie-seule-face-au-péril-libyen*.

38. In a study published by the Overseas Development Institute based on research conducted across the United Kingdom in 2012 has shown that British taxpayers after years of post-2008 austerity budgets are less enthusiastic about scaling up aid volumes, and calling for better ways to spend aid money. See Alex Glennie, Will Straw, and Leni Wild, "Understanding Public Attitudes to Aid and Development," London, UK: Institute for public Policy Research and Overseas Development Institute, June 2012, available from *www.odi.org/sites/odi.org.uk/files/odi-assets/publications-opinion-files/7708.pdf*.

39. George Parker and Stanly Pignal, "G8 Offers $20bn to Aid Arab Democracies," *Financial Times*, May 27, 2011, available from *www.ft.com/cms/s/0/aa9e92ca-884e-11e0-a1c3-00144feabdc0.html*.

40. Nadia Omrane, "Tunisie: Marzouki et Ben Jaâfar sous l'insulte, Jebali sous la coquette" ("Tunisia: Marzouki and Ben Jaafar Insulted, Jebali under the Duvet"), *Kapitalis*, December 19, 2012, available from *www.kapitalis.com/tribune/13456-tunisie-marzouki-et-ben-jaafar-sous-l-insulte-jebali-sous-la-couette.html*.

U.S. ARMY WAR COLLEGE

Major General William E. Rapp
Commandant

*****

STRATEGIC STUDIES INSTITUTE
and
U.S. ARMY WAR COLLEGE PRESS

Director
Professor Douglas C. Lovelace, Jr.

Director of Research
Dr. Steven K. Metz

Author
Dr. Mohammed El-Katiri

Editor for Production
Dr. James G. Pierce

Publications Assistant
Ms. Rita A. Rummel

*****

Composition
Mrs. Jennifer E. Nevil

Made in the USA
Monee, IL
04 November 2026

67380386R00036